The Food-Energy-Water Nexus 2016

Topical Conference at the 2016 AIChE Annual Meeting

San Francisco, California
13-18 November 2016

ISBN: 978-1-5108-3422-4

The Food-Energy-Water Nexus 2016

Topical Conference at the 2016 AIChE Annual Meeting

San Francisco, California
13-18 November 2016

AIChE
The Global Home of Chemical Engineers

Printed from e-media with permission by:

Curran Associates, Inc.
57 Morehouse Lane
Red Hook, NY 12571

CURRAN ASSOCIATES INC.
proceedings.com

Some format issues inherent in the e-media version may also appear in this print version.

Printed by Curran Associates, Inc. (2017)

For permission requests, please contact AIChE
at the address below.

AIChE
120 Wall Street, FL 23
New York, NY 10005-4020

Phone: (800) 242-4363
Fax: (203) 775-5177

www.aiche.org

Additional copies of this publication are available from:

Curran Associates, Inc.
57 Morehouse Lane
Red Hook, NY 12571 USA
Phone: 845-758-0400
Fax: 845-758-2633
Email: curran@proceedings.com
Web: www.proceedings.com

TABLE OF CONTENTS

476338 Review and Perspective on Nexus System Analysis

Abstract:

As the global population grows and increasingly lives in cities with lifestyles based on greater material consumption, more attention is being given to the integrated system that supplies our food, energy and water, the Nexus. The interaction between these three essential commodities can be seen in many ways. For example, production of sufficient food requires use of man-made fertilizer, to an extent that food and energy prices have become linked. Agriculture now accounts for 70% of water withdrawn from aquifers, streams and lakes. In many places fresh water is critically constrained, which affects our ability simultaneously to produce more food, grow crops for biofuels, and maintain natural ecosystems. And there is mounting concern about the effects on the Nexus of climate change, itself largely the result of fossil fuel consumption and land-use change (for agriculture).

Evidence for this growing interest in the Nexus is illustrated by a review of published literature (1). Over the 5-year period 2011-2016 the rate of publication of studies on the Food-Water-Energy Nexus in the scientific literature grew by a factor about ten. These publications underline the importance of Nexus concerns to policy-makers, transnational and regional institutions, and business/commerce. The study purposes of this body of literature can be summarized as

- reviews and Nexus issue discussion papers;
- quality-of-life studies to understand Nexus trade-offs or develop responses to achieve sustainability objectives (e.g., human well-being, healthy ecosystems);
- product studies to understand Nexus implications or develop responses related to providing products (e.g., a renewable energy technology, a service such as clean water); and
- studies to develop integrated system modelling tools.

Case studies tend to be of two types: those related to regional development, and those related to specific sectors. So Hoff's (2) background paper for the Bonn2011 conference outlines ten Nexus case studies, including Jordan and Ethiopia, the arid Ningxia region of China, and Gujarat. The Brazilian Cerrado savanna area offers a study in the competition between biofuel and food crops for land and water resources; the Nile Basin offers the same trade-off, with hydropower also playing a role (2). Responses by several companies to Nexus challenges in their own business areas are outlined in a report by Conservation International (3). These responses cover a wide range of policies and activities, but there are several common themes, including

- developing greater resource efficiency
- conducting resilience planning for price volatility and climate change
- systems thinking to design synergistic solutions (e.g., business decisions that also yield multiple environmental and social benefits)
- gathering new data to understand current and future trends
- stakeholder engagement to better understand the Nexus
- ecosystem valuation to help decision-making and balance trade-offs.

Particularly important in view of the trend of more people to live in cities are the (relatively few) studies of the "Urban Nexus", dealing with what may be termed the "metabolism" of the city. It is concluded that this metabolism, a hybrid of natural and artificial structures presents an opportunity for greater efficiency through better design and regulation. At present it is poorly understood, and studies are hampered by lack of data.

Case studies show the vital role played by stakeholder involvement in clarifying issues, priorities, and values. They also demonstrate the importance of an integrated systems view of the complex interrelationships of the Nexus when planning effective remedies. Assessments conclude that transformative social and political change is needed to create new structures, markets, and governance to deal with the Nexus if we are to meet agreed-upon sustainable development goals.

Needs for the future of Nexus studies include

- the publication of more case studies, particularly including stakeholder involvement;
- the creation of new frameworks and methodologies of analysis, including some that will integrate societal, environmental, and technical dimensions;
- the development of modelling tools accessible to a wider range of investigators (at present these tend to be highly specialized);
- the promotion of best practice in Nexus evaluation and effective use of study results;
- improved accessibility of Nexus data;
- metabolism modelling and engineering for the urban Nexus; and
- promotion and support from funding agencies and institutions for multidisciplinary and multinational studies of Nexus analysis, issues, and applications.

(1) Keairns DL, Darton RC, Irabien A 2016 ***The Energy-Water-Food Nexus*** Annu. Rev. Chem. Biomol. Eng. 7:239–62

(2) Hoff H. 2011. ***Understanding the Nexus. Background Paper for the Bonn2011 Conference: The Water, Energy and Food Security Nexus.*** Stockholm: Stockholm Environ. Inst.

(3) Gerholdt J, Pandya S. ***Resources: The Energy, Water, Food Nexus***. Arlington, VA: Conserv. Int. Bus. Sustain. Counc.

475423 Sustainability Concepts in the Food-Energy-Water Nexus: Chemical Engineering Perspective

Abstract:

The idea of sustainability and/or sustainable development emerged from the World Commission on Environment and Development (WCED), generally known as the Brundtland Commission. It defined sustainable development in "Our Common Future" as that which meets the needs of the present without compromising the ability of future generations to meet their own needs. Environmental sustainability and/or sustainable development "occurs when we maintain or improve the material and social conditions for human health and the environment over time without exceeding the ecological capabilities that support them". As such, sustainability has three underlying bases: (1) economic viability - the cost and business aspects of a project; (2) social concern - human health and social welfare; and (3) natural or ecological issues - depletion of natural capital and environment. Sustainable development has become a major driving initiative in engineering businesses throughout the world today. Our focus is on all aspects of sustainable development including environmental, industrial, business, economical, ethical, and social sustainability as they impact the availability of food, energy, and water. This is important because it ***has been generally acknowledged that the increasing use of non-renewable resources to support an increasing population has created an unsustainable or nearly unsustainable situation where the ability to provide adequate food, energy, and water is in question. Should this rate of use be maintained, two dire consequences will follow: first, the future generations will be unable to maintain a good standard of living, and second, developing countries will have less of an opportunity to bring their living standards to a level comparable to that of the affluent countries. This is important because the availability of food, energy, and water is an indispensable necessity for maintaining a good standard of living. The practice of chemical engineering, perhaps more than any other technical discipline, involves the use of both non-renewable and renewable materials, and energy and exergy (available energy) resources for the production of chemicals, value-added goods, services of commerce, and waste as***

pollution. This paper primarily deals with the ideas of environmental sustainability and/or sustainable development, methodologies to redesign sustainable chemical and allied processes and products development using the LCA approach. These principles are illustrated with a case study for the production of ethanol, an energy source, from molasses, a human food, in a combined agricultural and chemical process that requires substantial amounts of water.

461670 Optimizing Spatio-Temporal Sensor Placement for Nutrient Monitoring: Algorithmic Framework

Abstract:

INFEWS is a new initiative in NSF which deals with Innovations at the Nexus of the Food, Energy and Waters Systems. This project addresses the INFEWS: N/P/H2O solicitation. The food production system generates waste streams that are characterized by high concentrations of organic matter, nitrogen- and phosphorus-containing species in water. Therefore, monitoring nitrogen and phosphorous species is important for water quality requirements for agricultural as well as energy production. Currently these species are monitored via stationary monitoring stations. However nitrogen and phosphorous species move via agricultural run-off to other water systems, and requires portable sensors which can change the positions in real time. This type of dynamic sensing requires novel algorithms which decide sensor locations in real time in the face of inherent uncertainties in the fate and transport of the species. To develop such an algorithmic framework to solve the problem of sensor placement in real time is the objective of this proposal.

This paper presents the first part of the work which is the algorithmic framework for optimal spatial-temporal locations of the sensors in real time. The algorithmic framework is based on a novel algorithm called Better Optimization of Nonlinear Uncertain Systems (BONUS). We present a case study of an agent based model from social sciences for testing these framework. This model is based on behavior of cows in a field where there are grass patches. The cows move in a herd towards the greener pasture and eat grass till it finishes. The grass growth and cow movements are probabilistic. We are using four sensors for object tracking (e.g. cows) and these sensor move spatially and temporally in order to track maximum number of cows. Results show that the optimal movement of sensors allows to track maximum number of cows.

In the future, the algorithmic framework will be coupled with nutrient and fate transport models. A case study of a small watershed will be selected from literature for nutrient monitoring sensor location problem. Optimal spatial-temporal locations will be the result of this new framework for this case study.

Bandwidth Study on Energy Use and Potential Energy Saving Opportunities in Manufacturing Food and Beverages

Presenter:
Caroline Dollinger
Engineer, Energy Analysis
Energetics Incorporated

American Institute of Chemical Engineers
Annual Meeting
San Francisco, CA
November 14, 2016

Collaboration

Caroline Dollinger (Speaker)
Energetics Incorporated
cdollinger@energetics.com

Sabine Brueske
Energetics Incorporated
sbrueske@energetics.com

Energetics Incorporated is an energy consulting company, providing subject matter expertise to private and public sector clients

Joseph Cresko
U.S. Department of Energy
Advanced Manufacturing Office
joe.cresko@ee.doe.gov

Alberta Carpenter
National Renewable Energy Laboratory
Alberta.carpenter@nrel.gov

2

Overview

- An *Energy Bandwidth Study* is currently being performed to analyze energy use and energy saving opportunities in energy intensive manufacturing sectors
- This presentation focuses on the work being done for a *Food and Beverage Manufacturing Energy Bandwidth Study*
- Energy data will broken down by sub-sector, product, and unit operation to identify processes and energy-saving technologies with the greatest opportunities for energy savings

Results presented in this talk are draft data; the work is still ongoing and a final report is expected to be published next year.

3

ENERGETICS
A Subsidiary of VSE Corporation

Energy bandwidth studies can be a useful tool for assessing energy savings opportunities in manufacturing

Energy bandwidth studies frame the range (or *bandwidth*) of potential energy savings in manufacturing, and technology opportunities to realize those savings.

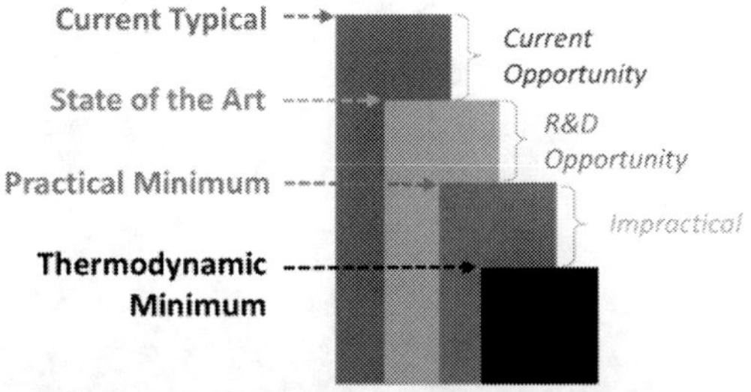

Measures of energy intensity studied:

Current Typical (CT)	State of the Art (SOA)	Practical Minimum (PM)	Thermodynamic Minimum (TM)
Determined from a literature review and stakeholder outreach, based on current typical manufacturing processes in the U.S.	Determined from a literature review and stakeholder outreach, based on the most energy-efficient technologies and practices available worldwide	Modeled based on plausible energy savings from identified R&D technologies under development worldwide	Calculated analytically using a Gibbs free energy approach assuming ideal conditions

4

ENERGETICS
A Subsidiary of VSE Corporation

Six new energy bandwidth studies on lightweight materials were recently completed in draft form

- **These studies have previously provide a deep-dive analysis into manufacturing energy use and potential energy savings for:**
  - **Four energy-intensive manufacturing sectors**
  - **Manufacturing of six lightweight materials**
- **This talk will focus primarily on a new bandwidth on food and beverage manufacturing.**
- **Published bandwidths (including this one, once completed) are available on the DOE Advanced Manufacturing Office website**

Summary thumbnail for bandwidth study on petroleum refining

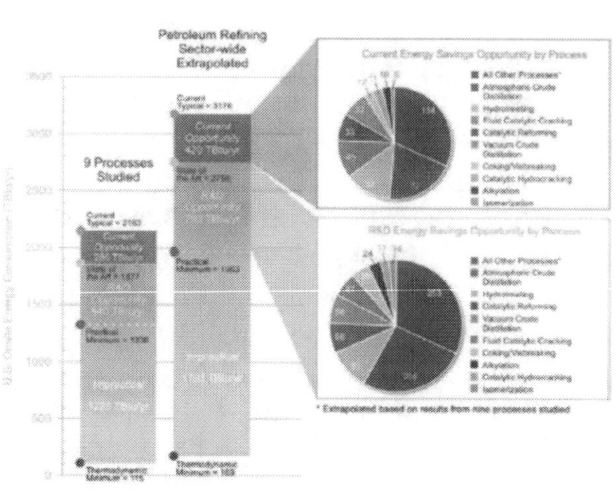

5

ENERGETICS
A Subsidiary of VSE Corporation

First Four Industry Bandwidths – US Results

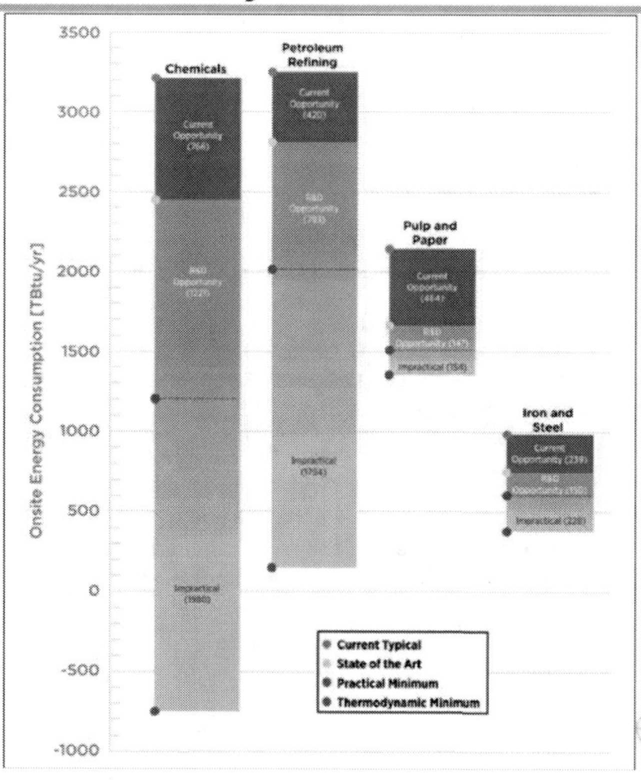

ENERGETICS
A Subsidiary of VSE Corporation

Challenges in Researching Energy Use and Savings in the Sector

- **Highly diversified sector that produces thousands of different products**
- **Processing facilities range from small to large units, normally producing more than one product**
  - **As a result energy intensities will vary from facility to facility and from product to product**
- **Less active in the R&D, with a focus on energy savings technologies, compared to other sectors**
- **Competitive marketplace, where energy efficiency may help to offer a competitive advantage**
- **Largely mechanized processes, difficult to ascertain and source energy data at the sub-process level**

7

ENERGETICS
A Subsidiary of VSE Corporation

Food & Beverage Manufacturing Bandwidth
Study Coverage

Food and Beverage Sub-sectors Studied	Food and Beverage Products Studied
Grain and Oilseed Milling	Wet corn milling- corn oil and by-products Soybean oil and meal Other vegetable oils and meals
Sugar Manufacturing	Beet sugar Cane sugar
Fruit and Vegetable Preserving and Specialty Food Manufacturing	Frozen fruits, vegetables, juice Canned fruits, vegetables Pickled vegetables Dried and dehydrated fruits and vegetables
Dairy Product Manufacturing	Milk Butter Cheese Dry, condensed, evaporated milk Ice cream/frozen desserts
Animal Slaughtering and Processing	Beef, pork, chicken, lamb Animal fat, lard, edible tallow
Beverage Manufacturing	Soft drinks Beer, wine, distilled liquors

8

ENERGETICS
A Subsidiary of VSE Corporation

Food & Beverage Manufacturing Bandwidth
Study Coverage

Food and Beverage Sub-sectors Studied	Energy Intensive Sub-processes
Grain and Oilseed Milling	Crushing Extraction Evaporation Dewatering and drying
Sugar Manufacturing	Dewatering and drying Evaporation Vacuum pans (crystallization)
Fruit and Vegetable Preserving and Specialty Food Manufacturing	Heat sterilization/retorting Cooking/blanching Freezing Concentration Evaporation Drying Canning
Dairy Product Manufacturing	Pasteurization Freezing Concentration
Animal Slaughtering and Processing	Rendering
Beverage Manufacturing	Beer processes (brewhouse, packaging) Wine processes (refrigeration)

Food and Beverage Manufacturing Bandwidth
Scope Description

What's Out- Agriculture Sector NAICS 11

Establishments primarily engaged in growing crops, raising animals, from a farm, ranch, or their natural habitats

111-Crop production

112-Animal production

What's In- Manufacturing Sector NAICS 31

Industries in the Food Manufacturing subsector transform livestock and agricultural products into products for intermediate or final consumption (basically, any food that is processed)

311 – Food manufacturing

What's Out- NAICS 42, 44, 72

424 – Food Wholesalers

445 – Food and Beverage Stores

722 – Food Services and Drinking Places

ENERGETICS
A Subsidiary of VSE Corporation

10

Food and Beverage Sector

U.S. Energy Consumption Overview

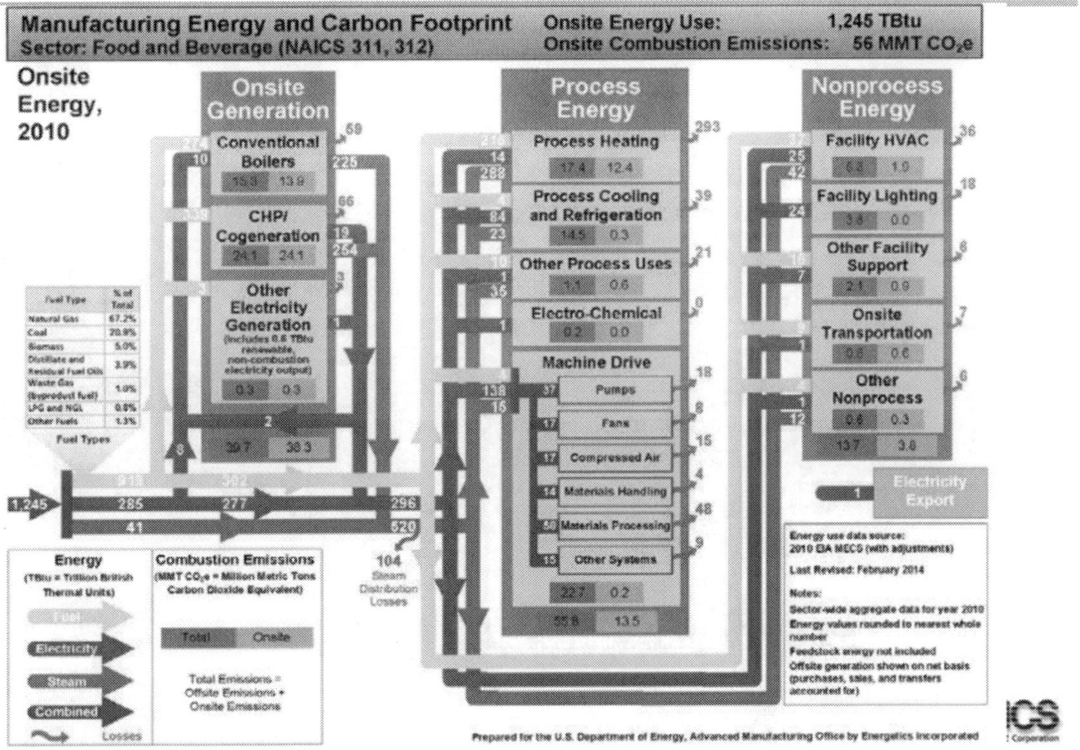

Food and Beverage Bandwidth

U.S. Baseline Energy Consumption, 2010

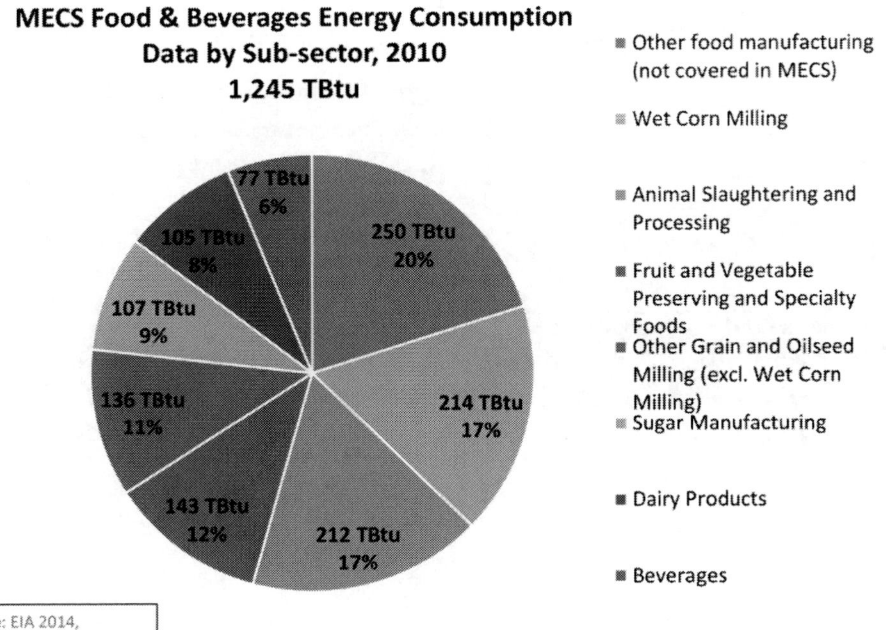

Source: EIA 2014, Manufacturing Energy Consumption Survey (MECS)

12

ENERGETICS
A Subsidiary of VSE Corporation

Food and Beverage Bandwidth
U.S. Production for Sub-sectors Studied, 2010

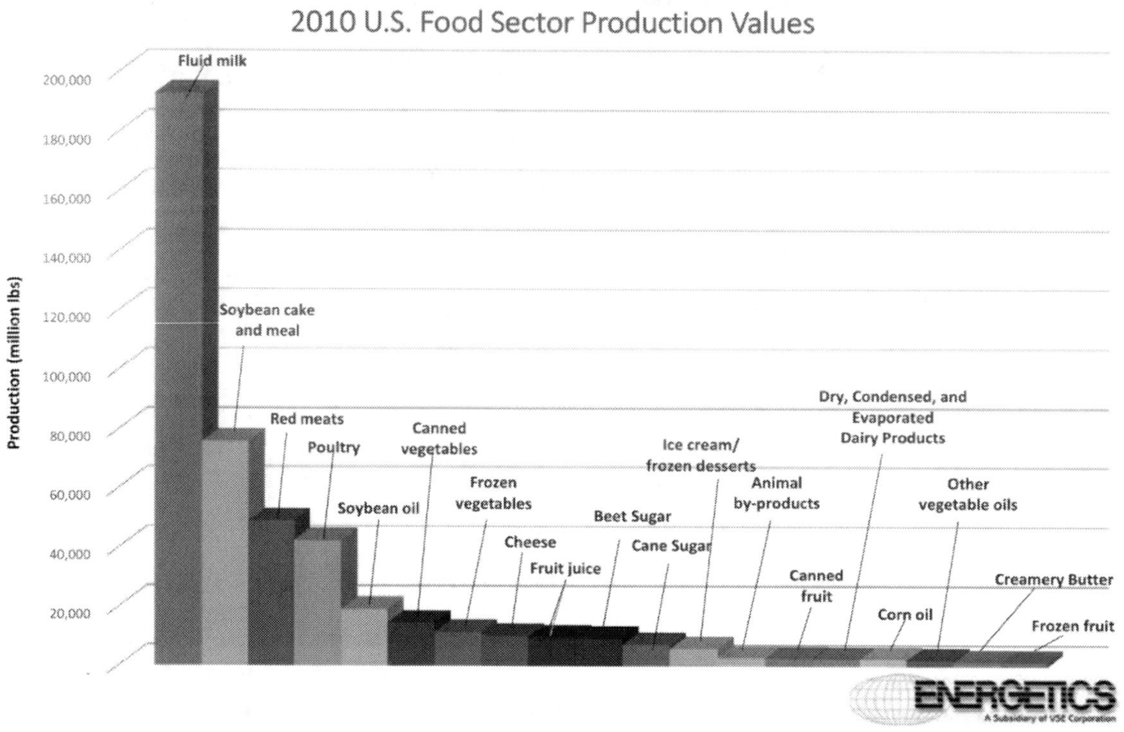

Dairy Products Manufacturing
Energy intensive sub-processes being analyzed

- Pasteurization
- Cooling/refrigeration
- Concentration
- Freezing
- Motors/pumps
- Spray Drying
- Canning

Manufacturing Process	CT (TBtu/year)	% of sub-sector
▶ **Pasteurization**	20	18.6%
▶ **Cooling**	16	15.6%
▶ **Motors, pumps**	9	8.4%

Grain and Oilseed Milling
Energy intensive sub-processes being analyzed

Wet Corn Milling

- Starch dewatering and drying
- Gluten thickening and drying
- Gluten feed drying
- Steeping
- Steepwater evaporation
- Fiber and gluten recovery
- Germ dewatering and drying
- Grinding

Other vegetable oils/meals

- Crushing and extraction
- Neutralization
- Oil refining
- Deodorization

15

Fruit and Vegetable Preserving
Energy intensive sub-processes being analyzed

Canned Fruits & Vegetables

- Washing
- Retorting
- Blanching/cooking
- Canning

Frozen Fruits & Vegetables

- Washing
- Peeling
- Blanching/cooking
- Freezing

Other fruit/ vegetable products

- Pasteurization (fruit juice)
- Dehydrating
- Retorting (fruit juice)

16

Animal Slaughtering and Processing
Energy intensive sub-processes being analyzed

Red Meat Products

- Dressing and cutting
- Refrigeration
- Processing (curing, smoking, cooking)
- Packaging
- Splitting
- Scalding
- Singeing
- Blood processing
- Edible rendering

Poultry Products

- Scalding, slaughtering, defeathering
- Evisceration and cooling
- Refrigeration
- Liquid effluent treatment
- Packaging

17

Current Typical (CT) energy intensities and consumption for red meat products, U.S. (2010)

Manufacturing Process	CT (Btu/lb)	TBtu/year (% of sub-sector)
Chilling	468	23 (10.8%)
Processing (curing, smoking, cooking)	320	16 (7.4%)
Edible rendering	4,573	14 (6.5%)
Dressing and cutting	181	9 (4.2%)
Packaging	93	5 (2.1%)
Scalding*	149	3 (1.6%)
Splitting*	117	3 (1.2%)
Singeing*	82	2 (0.9%)

* Pork products

18

ENERGETICS
A Subsidiary of VSE Corporation

Animal Slaughtering and Processing

Example of challenges in data and analysis

- **Lack of recent analysis done on energy usage in U.S. manufacturing facilities, must rely on internationally available data**
- **Most estimates of energy intensity are presented in terms of total energy/product or input (e.g., kWh/tonne), not broken down by sub-process**
- **Studies done on individual plants may not be applicable to be generalized sector-wide**
- **Energy data presented in different forms (consumption per head, carcass weight, dressed weight, product weight)**
  - **Involves accurate conversions to calculate total energy consumption using production values from USDA, other assumed factors**
- **If data is provided for sub-processes, only provided for a few, not allowing for calculation of balance**

ENERGETICS

A Subsidiary of VSE Corporation

Current Opportunity and *R&D Opportunity*

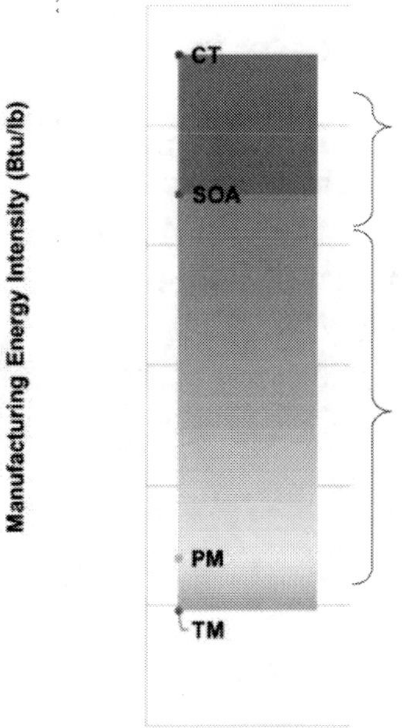

Current Opportunity

Energy savings if the best technologies and practices available were used to upgrade production

R&D Opportunity

Additional energy savings if applied R&D technologies under development worldwide were successfully deployed

<u>SOA Technology Examples:</u>

- **Steam system improvements**
- **Adjustable-speed drives**
- **Air compression improvements (reduce leaks, efficient vacuum pumps)**
- **Alternative technologies**
- **...**

<u>R&D Technology Examples:</u>

- **Membrane filtration**
- **Microwave heating**
- **Advanced Ohmic Heating**
- **High pressure processing**
- **Cost efficient thermal management in motors using nanomaterials**
- **...**

ENERGETICS

A Subsidiary of VSE Corporation

What about the Water part of the Nexus?

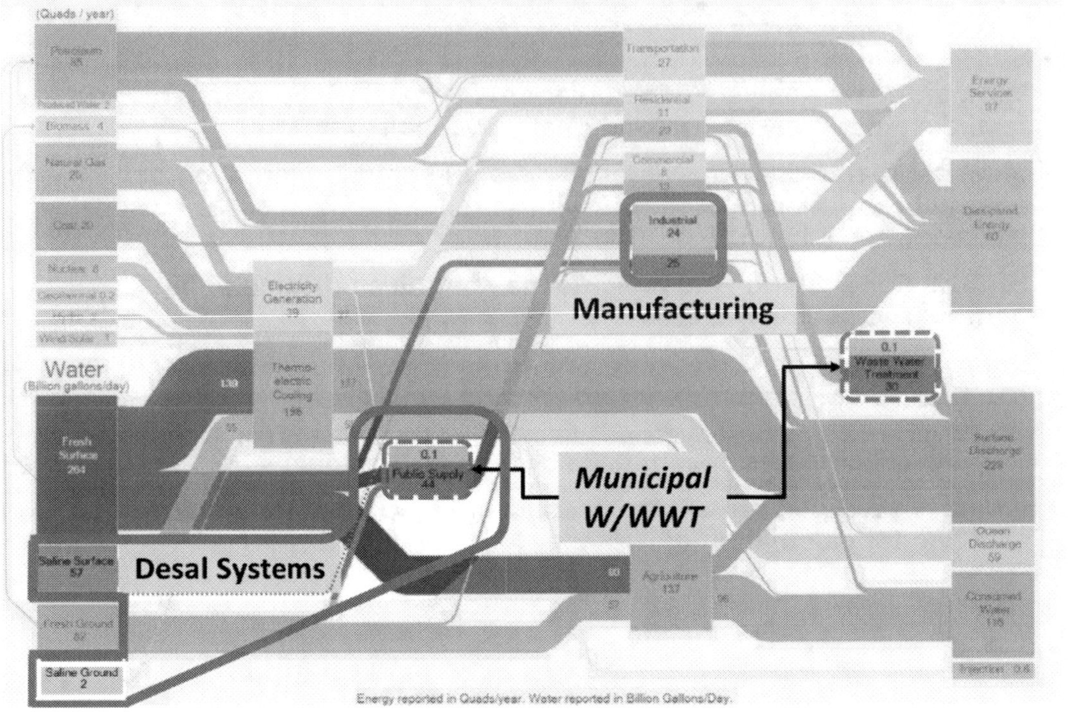

Conclusions

- An *Energy Bandwidth Study* is being performed to analyze manufacturing energy use and energy saving opportunities for the food and beverage manufacturing industry
- Manufacturing energy data are broken down by sub-sector, product, and sub-process to identify processes and energy-saving technologies with the greatest opportunities for energy savings
- This type of study has not been done in the past, so data must been gathered carefully; also there is a lack of data in certain areas
- While more research is to be done, research to date implies that there may be more current rather than future energy savings opportunity, due to the fact that there are few areas with active R&D in food and beverage, while more traditional areas for energy efficiency savings (steam systems, etc.)

Results presented in this talk are draft data; studies are currently being peer reviewed. If you are interested in participating, please contact me!

22

ENERGETICS
A Subsidiary of VSE Corporation

Thank you!

Energetics Incorporated

Caroline Dollinger

(CDollinger@energetics.com)

ENERGETICS

A Subsidiary of VSE Corporation

23

468656 Insight-Based Design of Local Integrated Systems for Food, Energy and Water

Abstract:

The importance of Food-Energy-Water nexus (also referred to as Water-Energy-Food nexus) has been widely recognised by national governmental bodies and international organisations such as the United Nations, for its significance to sustainable development and national security. Though these three sectors are inextricably linked, as actions in one sector have impacts in one or both of the others, these areas have too often been considered individually. Thus, there is a need to look at them from a holistic perspective and develop tools that take into consideration their interdependencies. This opens up opportunities for process system engineering research. As such, the aim of this study is to develop a systematic insight-based approach for the design of an integrated local production system particularly to meet local demands for food, energy and water. One of the benefits of the insight-based approach is that it allows decision variables to remain tractable throughout the design process. Unlike mathematical optimisation approaches, intermediate results from insight-based approaches can be analysed, interpreted and used to inform the implications of a given design for the operation of the whole system, thus offering a more practical tool for decision-makers and local planners to use. Insight-based approaches which include techniques such as pinch analysis for heat integration and mass integration have proved particularly useful for the integration of energy and material streams in industrial production processes within eco-industrial parks and industrial symbiosis systems. They have been successfully implemented for enhancing resource efficiency and sustainability of processes involving heat, water and hydrogen, to name a few applications. Although the proposed study also pursues higher resource efficiencies through integration of resources, the design of the local integrated production system has a distinctive emphasis on local resources and demands and on the holistic consideration of all types of agricultural, industrial and municipal processes to take place at the locality of concern. This work will combine existing methods and adapt them to enable the identification of the efficient use of various locally available resources. A net gain indicator will be defined to quantify the avoided cumulative effort that would have been spent for importing resources to satisfy the demand.

Using the net gain indicator, parallel processing and utilisation options within and between different subsystems can be compared and the optimal actions identified. The proposed insight-based approach for the design of local integrated system of food, energy and water will be illustrated through a case study based on an eco-town in the UK.

468765 Recovering Runoff Particulate-Bound Phosphorus Via Fungal Bioextraction

Abstract:

Phosphorus (P) loss is a key issue affecting food-energy-water nexus sustainability. P in fertilizers is critical for ensuring food security for the continuously growing world population. Currently, P is obtained via energy-intensive mining and processing of phosphate rock, which is becoming more scarce and concentrated in a few geographical areas. Excessive fertilizer application due to slow diffusion and heavy fixation of P in soils have led to massive P losses amounting to almost half of all mined P due to agricultural soil erosion and runoff. P-rich runoff streams discharging into surface water bodies can cause eutrophication, toxic algal blooms, and overall quality deterioration of freshwater resources. In this work, we propose capturing and processing runoff sediments using a low-energy and inherently benign fungal bioprocessing method to unlock a previously untappped source of recoverable phosphorus for nutrient recycling and closing the loop of P use. This can potentially reduce the demand for mined sources while at the same time help protect watersheds from the devastating effects of eutrophication.

This paper presents the preliminary findings of our work involving the use of phosphorus-solubilizing fungi (PSF) for extracting and recovering particulate-bound phosphorus in erosion and runoff streams. PSF species can generate organic acids and phosphatase enzymes that can induce mobilization of matrix-associated or surface-adsorbed phosphate ions on inorganic minerals in runoff particulates, and degrade phosphate ester bonds in sedimentary organic phosphorus compounds to facilitate the release of phosphate ions, respectively. Samples of runoff sediments discharging into a eutrophic lake in Kalamazoo, MI were collected and characterized for total P content and P speciation. P extraction experiments on runoff sediment samples were performed using model organic acids and spent culture broth *Aspergillus niger* ATCC 15475. Phosphate solubilization and changes in P speciation of the runoff sediments during treatment were measured using a sequential chemical extraction method and colorimetric analysis. Changes in runoff sediment mineralogy was analyzed using X-ray Diffraction (XRD) spectroscopy. The results indicate that oxalic and citric acids can extract and solubilize a significant fraction of phosphates bound to the iron, aluminum, and calcium mineral components of the runoff sediments. A kinetic model describing phosphate mobilization via ligand-exchange and induced desorption mechanisms was developed and fitted into the experimental data. We also observed a significant reduction in the organic P fraction of sediments treated with spent fungal culture broth, indicating the possibility of phosphatase enzyme production and activity. Thus, changes in organic P speciation in the sediments were monitored using solution ^{31}P nuclear magnetic resonance (NMR) spectroscopy. The findings demonstrate the feasibility of the proposed technology for P recovery from nonpoint nutrient pollution sources such as runoff sediments. It will help develop an effective, efficient, economical, and environmentally benign system that will benefit the community, agriculture and food production systems, and the environment in terms of resource security and environmental sustainability.

469658 Preliminary Study on Detroit's Urban Food-Energy-Water (FEW) Nexus

Abstract:

With 66% of the worldâ€™s population expected to live in urban areas by 2050, cities around the world will face numerous challenges in sustainable provision of food, energy, and water. These challenges are particularly complex due to the interconnectedness of the food system, energy system, and water system. For example, water is required for food and energy production; energy is needed for food production as well as water and wastewater treatment and distribution; and industrial agriculture and food production lead to oversupply of nitrogen and phosphorus in the urban water system through fertilizer use. Therefore, policy and technology solutions addressing urban food-energy-water (FEW) challenges need to be evaluated through the lens of FEW nexus to identify co-benefits and avoid unintended consequences. This calls for innovative approaches to model the urban FEW nexus as an integrated whole, instead of examining them individually.

Existing studies primarily investigated urban FEW systems individually (e.g., nutrient flows, energy flows, and water flows) or focuses on two of the three FEW systems (e.g., water-energy nexus and food-water nexus). Existing studies that consider all FEW systems primarily aim to understand the impact of one system on the other two, such as how urban water-sector innovations change food and energy flows. There lacks an integrated approach that characterizes the comprehensive interdependence of urban FEW systems.

In this study we developed a systems modeling framework for the urban FEW nexus. We also applied this framework to the Detroit Metropolitan Area (DMA) as a preliminary study. The urban FEW systems modeling framework we proposed is based on material and energy flow analysis (MEFA) which quantifies flows and stocks of a group of substances within a system during a given period of time. This framework includes three steps:

Step 1: Individual FEW systems are modeled as resource flows across major sectors (e.g., water treatment, wastewater treatment, food production) of the urban system. Given the variety of food crops and products, we used nitrogen and phosphorus flows as the currency to represent the food system. Quantities of resource flows (water, energy, nitrogen, and phosphorus) are estimated based on a variety of data such as population, water withdrawal, and wastewater generation that were collected for the DMA case study.

Step 2: We identified and measured interdependence of individual FEW systems. This was done through literature review for common FEW interactions and quantitative estimation of resource flows between individual FEW systems (e.g., the amount of water used for food production and energy production).

Step 3: Each individual FEW system is characterized as a network of resource flows. Based on the interconnections between individual resource flow networks obtained from Step 2, we integrated the individual FEW resource flow networks as a network of four networks (water, energy, nitrogen, and phosphorus). This provides a mathematic representation of the urban FEW nexus, which will be the base of further analysis and modeling.

474593 Innovations at the Nexus of Food, Energy, and Water Systems (INFEWS): A National Science Foundation (NSF) Initiative to Examine Critical Issues in Today's World

Abstract:

At the end of 2015, the National Science Foundation (NSF) put out its first request for proposals for the INFEWS program (Innovations at the Nexus of Food, Energy, and Water Systems). Over the next five years the program plans to fund interdisciplinary projects that fall into four tracks 1. System Modeling, 2. Visualization and Decision Support, 3. Innovative System Solutions, and 4. Education and Workforce Development. The program expects to tackle Food-Energy-Water System (FEWS) research through integration across disciplines, in which agricultural and food systems research is a primary component. To this end NSF has partnered with the United States Department of Agriculture (USDA) to fund related research. This program developed from the Science, Engineering and Education for Sustainability (SEES) program, which in 2015 funded 17 FEWS workshops. These interdisciplinary workshops brought together researchers and other stakeholders from universities, local, state, and federal governments, industry, and non-profits to examine a wide range of FEW questions and identify the research, data and technological needs to solve them. This presentation will cover an overview of the development of the INFEWS program, some pressing FEW needs as identified by the broader research community, and national and international opportunities for future engagement with the program.

474707 The “Root” of the Nexus: Soil-Based Biotechnology for Sustainable Agriculture

Abstract:

Food, energy, and water nexus research by chemical engineers is typically focused on the built environment. For example, innovations in sustainable food production might offer food processing, distribution, or waste recycling technology that minimizes water or energy use. However, recent technological advances in different fields ranging from materials and manufacturing to molecular tools and systems biology are also enabling more sustainable primary production of food in the natural environment. Most of our food is produced in soil, a dynamic natural system of enormous complexity. In particular, the rhizosphere (the region of soil adjacent to plant roots) is a site of steep chemical gradients, plant growth-limiting multiphase transport phenomena, and dynamic inter-kingdom collaboration among members of the root microbiome. Emerging technology is re-engineering the root zone for more sustainable primary food production. Examples of emerging agriculture biotechnology to produce more food with less water and energy include: (i) advanced seed coating technology and nano-enabled agrochemical delivery systems; (ii) plant-growth promoting bacteria and microbial products to improve plant nutrition and enhance moisture retention in soil; and (iii) sensors and sensor networks for precision agriculture. By learning from and working with natural systems, chemical engineers are developing innovative agriculture biotechnology for more sustainable primary food production.

474596 The Role of Waste Streams in the Energy-Water Nexus

Abstract:

The energy-water nexus has helped focus attention on the water used for energy production and energy used in providing water services. However, there has been a lack of attention to the downstream aspects of this nexus; that is, impacts of related waste streams on land, water and air. Primary technology choices (e.g., electricity generation technology, drinking water source) result in differing waste streams which in turn require decisions on the level of treatment and disposition of the related waste products. Each of these choices have related implications for energy use, water use, environmental impact and system vulnerability (e.g., power plant thermal discharge puts operations at risk to drought). Remarks in this panel will endeavor to establish a framework and context for a more holistic treatment of waste streams within the broader energy-water nexus.

474597 Process Development Challenges and Opportunities in Food-Energy-Water (FEW) Nexus

Abstract:

Presentation will start out with a high level view of Food-Energy-Water (FEW) nexus for illustrating complexity of the topic. Global changes in population, energy needs, and water use will drive technological innovations and rapid commercialization needing chemical engineers involved in process development. Connections and interdependencies discussion among these three systems will illustrate how engineering solutions from fields like systems approach, catalysis, biocatalysis, environmental sustainability, etc. will contribute to better understanding the FEW nexus challenges and successfully overcoming them.

Broad overview of current and future process technologies important to FEW nexus selected from various NSF and AIChE workshops and recent developments reported in literature will be presented. Challenges and opportunities for successful transition from concepts to commercialization of such technologies will be discussed with actual examples at various stages of development. Examples will be presented wherever possible to show how process research and innovation, pilot plants, technology transfer, process intensification and micro process engineering as well as product design will contribute to the nexus. Need for collaborative efforts among academics, government labs, and industry across multiple disciplines and countries for local applications for ultimate success in the FEW nexus will be highlighted.

The talk will also cover selected areas where chemical engineers in general and process development engineers in particular can make significant contributions. Critical need for process research, development, and demonstrations that enables new means of adapting to future challenges will be presented. As an example, out of the three main factors nitrogen, phosphorous, and water that limit our ability to produce enough food, nitrogen and water will be covered in detail. This section of the presentation will show how catalytic and waste water treatment technologies will play the

role in connection to the FEW nexus.

Main goal of the presentation will be to make audience aware of theoretical and experimental process development challenges and opportunities of the future.

474595 Systems Engineering Methods for Food-Water-Energy Nexus

Abstract:

As the global population grows, consumption of water, energy, and food will also increase, placing stresses on these three sectors, raising the importance of the Food-Water-Energy Nexus (FWEN). This article highlights research challenges and identifies process systems engineering research opportunities to appropriately model and optimize the WEFN. A brief overview of relevant, foundational FWEN research is first presented. We then identify challenges in the multiple scales of the FWEN, ranging from the household scale to the global scale. There are further challenges with appropriate system boundary definitions, and challenges in modeling the decision-making and conflicting objectives of multiple stakeholders in the FWEN. Uncertainties of all kinds appear at all scales of the FWEN and must also be considered. We use two motivating FWEN examples to frame these challenges and propose future avenues and opportunities for research. Possible approaches to the aforementioned challenges are proposed.

466874 Managing Trade-Offs Between Food, Renewable Energy and Ecosystem Services

Abstract:

Competition between renewable energy and food production, driven by increasing food and energy demands, creates incentives to increase agricultural productivity. In the short term, productivity can be increased through the use of intensive farming management practices that use extensive tillage and large quantities of chemical fertilizers. Such practices have been shown to contribute to ecological degradation of farmland via an increase in soil erosion and a decrease in soil carbon, among other impacts. The surrounding watershed is also affected negatively by the nitrate and phosphorus-contaminated runoff from fertilized fields, which degrades the water quality and contributes to the growth of harmful cyanobacteria and other microorganisms. In the long term, this ecological degradation results in reduced agricultural productivity, such that increasing amounts of chemical fertilizers and other technological inputs are required. Farming for maximum productivity without accounting for the ecological impacts thus creates a vicious cycle, in which productivity is maintained by inflicting ecological harm.

Experimental studies have demonstrated that conservation farming practices, with reduced tillage and in some cases reduced amounts of fertilizers, insecticides and herbicides, can mitigate the ecological damage of conventional agriculture. Farming with reduced or no tillage helps to increase and maintain the carbon content in soil, providing an offset for the emissions produced in the agricultural life cycle, and reduces the rate at which soil erodes. Reducing chemical fertilizers or partially replacing them with soil amendments such as agricultural residues reduces the concentration of nitrates and phosphates in contaminated runoff and cuts down on wasted fertilizer as more of the applied fertilizer is taken up by crops. Conservation practices offer myriad ecological benefits but do not necessarily offer similar productivity benefits: the rate at which biomass is produced can be decreased under conservation farming, creating trade-offs between ecological preservation, agricultural production, and farm economics. With demand for both food and energy likely to increase for the foreseeable future, the need for food and bioenergy crops must be balanced against the need for ecological preservation.

The objective of this work is to design a system for the co-production of food and renewable energy with respect to three conflicting criteria: food productivity, energy productivity and net ecosystem service (ESS) supply. Ecosystem services are the goods (food, fuel, materials) and services (climate regulation, water quality regulation, erosion control) provided by nature that are necessary to support human activities, including agriculture. Here we define net ESS supply as the difference between ESS supply, provided by nature, and ESS demand, required by human activity. Climate regulation is a simple example. The net climate regulation supply is the quantity of greenhouse gas emissions generated by human activities, less the quantity of emissions that can be sequestered or otherwise removed from the atmosphere by ecological processes such as biomass growth and sequestration in soil. Net ESS supply provides a quantitative way to measure ecological degradation: a negative net supply means that the demand for ESS is greater than can be supplied by nature, and ecological degradation is the likely consequence. In this work, we will identify options for food and renewable energy co-production that have a net ESS supply greater than or equal to zero, for key services.

The system we consider is comprised of two stages: land use and biomass-to-energy conversion. Land use includes several options for crops both grain and cellulosic, rotation systems and farming management practices, as well as the production of wind and/or solar electricity. Land can also be used to create additional ESS supply through reforestation; however, land used for ESS supply cannot also provide food or energy, thus there is competition between the land uses. Options for biomass-to-energy conversions cover a range of liquid fuels, both ethanol and hydrocarbon, as well as electricity. Grain crops can be sold directly as food products or converted to liquid fuels; the system thus has a variety of configurations in which both food and renewable energy is produced from the same farmland. The land use and biomass-to-energy conversions are modeled at the value chain and equipment scales, respectively, of the process to planet (P2P) multi-scale modeling framework. For this system, the value chain scale may be thought of as a regional scale. The third and largest scale of the P2P framework, the economy scale, is used to capture the life cycle of all inputs to the land use and biomass-to-energy conversion stages. Net supplies of two ESSs are quantified as decision criteria: climate regulation, quantified with carbon dioxide emissions and carbon sequestration, and air quality regulation, quantified with nitrogen oxide emissions and removal. Carbon dioxide emissions (climate regulation demand) is quantified at all scales, equipment, regional and life cycle. Carbon sequestration (climate regulation supply) is quantified at the regional scale only, as the land use options are the only components of the system that provide a variable supply of ESS. Air quality regulation is quantified only at the smaller two scales, with supply again provided only by the land use options and demand created by both the land use options and the biomass-to-energy conversion options. Air quality regulation is not considered at the life cycle scale because the spatial presence of criteria air pollutants and its effects, are either local or regional in nature depending on the pollutant molecule.

To analyze trade-offs between food, energy and ecosystem services, the co-production system is optimized under several constraints. Food and energy production are constrained relative to the maximum production possible, to ensure that all optimal system designs are truly co-production systems. Where possible, net ESS supply is also constrained to be greater than or equal to zero; this is not always achievable for air quality regulation, but net climate regulation supply can reach above zero for many system configurations. Within these constraints, the system is optimized for maximum food production, maximum energy production, and maximum net ESS supply for both services. With the design space thus defined, Pareto curves are found between food production and net ESS supply, and between energy production and net ESS supply, in order to identify sets of compromise system designs. Preliminary results indicate that maximizing either food or energy productivity leads to the worst possible performance for net ESS supply, as expected. However, by either sacrificing some amount of productivity or by using land for both food and energy production and ESS supply, designs can combine adequate net ESS supply with relatively high food and energy productivity.

461799 Efficient Solar Thermal Hydrogen, Electricity and Fresh Water Coproduction Process Synthesis

Abstract:

Global population growth, increase in per capita GDP and change in consumption habits create increasing pressure on food, energy and water (FEW) supply. For the past century, with lower FEW needs and abundant fossil resources, energy, chemical, fertilizer and even fresh water supply via desalination have been predominantly dependent on limited fossil resources. However, concerns regarding limited fossil fuel reserves and increasing greenhouse gas (GHG) emissions from fossil fuels accelerate advancements and implementation of renewable energy conversion processes. Among these alternatives, solar energy conversion processes are particularly prominent because of sunâ€™s tremendous potential [1]. Yet intermittencies and land availability constraints for solar energy collection are the grand challenges for solar energy conversion technologies and demand high conversion efficiency and synergistically integrated energy storage solutions [2-3]. Recently, it has been shown that hydricity concept, synergistic solar thermal hydrogen and electricity coproduction, presents a potential solution for continuous and efficient power supply and also an exciting opportunity to envision and create a sustainable economy to meet all the human needsâ€”namely, food, chemicals, transportation, heating, and electricity [1].

Here, we introduce an integration strategy for the coproduction of fresh water, hydrogen and electricity from concentrated solar thermal energy and present potential impacts of the proposed system on FEW nexus. The integration of solar thermal power production and solar thermal hydrogen production techniques reduces the exergy losses associated with each one of the systems and thus provide an efficient solution. Solar conversion processes are evaluated based on the process *exergy efficiency* that refers to the fraction of incident solar exergy that is directly recovered as the net exergy output, which is defined as the sum of electricity and the hydrogen exergy output.

The proposed efficient solar trigeneration process presents: i) a continuous power and fresh water supply solution that

can achieve high solar conversion efficiencies when combined with the proper hydrogen power cycle and ii) numerous opportunities in chemical industry, transportation sector and in particular in food production via continuous supply of fresh water for irrigation and fertilizer production from coproduced hydrogen. Coproduced products can supply all FEW needs especially in the case of production of fertilizers from the coproduced hydrogen. [1] Gençer E, Mallapragada DS, Marechal F, Tawarmalani M, Agrawal R. Round-the-clock power supply and a sustainable economy via synergistic integration of solar thermal power and hydrogen processes, Proceedings of the National Academy of Sciences (PNAS), 112(52), 15821-15826, 2015.

[2] Gençer E, Al-musleh E, Mallapragada D, Agrawal R. Uninterrupted Renewable Power through Chemical Storage Cycles. Current Opinion in Chemical Engineering, 5, 29-36, 2014.

[3] Gençer E, Agrawal R. A commentary on the US policies for efficient large scale renewable energy storage systems: Focus on carbon storage cycles, Energy Policy, 88, 477-484, 2016.

465212 Food-Energy-Water Nexus: Modeling Energy and GHG Emissions of Water Embodied in U.S. Domestic Food Transfers

Abstract:

Global population growth, climate change, and increasing urbanization are pressurizing already constrained essential food, energy, and water resources. Food, energy and water (FEW) systems are deeply interwoven, interconnected, and necessitate an integrated management approach. Central to understanding the complexity of the FEW nexus is quantitatively mapping and understanding their complex interdependencies. Existing studies have focused mostly on a single or two dimensions of the FEW nexus. There is a critical need for a holistic systems-level approach for studying FEW systems in a joint manner as opposed to studying them in isolation to avoid unintended consequences and encourage optimal outcomes. We integrate concepts of network theory with life cycle information to develop an understanding of the domestic FEW nexus for the United States.

Using publicly available disparate datasets, we develop a weighted and directed network model of interstate food transfers for the United States. The interstate food transfer model is translated into networks of embedded irrigation water, energy, and greenhouse gas (GHG) emission flows utilizing information on water footprints, irrigation datasets, and life cycle inventory databases. Specifically, we focus on four food commodity groups: livestock, cereal grains, meat, and milled grains. We utilize network theory tools and metrics to understand the structure, robustness, and environmental sustainability of these networks. Preliminary results indicate that over 600 million tons of food commodities were transferred across the U.S. along with 230 billion m^3 of virtual irrigation water. Additionally, 450 billion MegaJoules (MJs) of primary energy and 30 billion kg of CO_2-equivalent emissions were embedded in virtual irrigation water transferred across the U.S. It is observed that livestock and meat contributed only 13% by mass but accounted for 60% of embodied water, energy, and GHG transfers compared to grain based commodities. From a network perspective, the unweighted food transfer network is well-connected with majority of states participating in high volume of trade. However, the weighted network structure reveals that the majority of the food flow and embodied resources and emissions are controlled by a few states. Furthermore, meat and milled grain networks are denser compared to livestock and cereal grains networks. A robustness analysis highlights that despite the presence of vulnerable key states, the network is fairly robust to both random and targeted disruptions. These results reveal that while the domestic food transfer network is robust against extreme disruptions on specific nodes, the dense food transfer patterns of resource intensive products have important implications for sustainability of the food network.

465536 Water Footprint of Hydrotreated Renewable Jet Fuel Produced through Rapeseed Rotation with Wheat and Other Crops in North Dakota

Abstract:

Hydrotreated Renewable Jet (HRJ) fuel from oilseed plants has increasingly become important for the aviation sector to address energy security and climate change mitigation. Rapeseed is considered a promising HRJ source for its high quality oil content and the potentially attractive agro-economic benefit to replace the fallow period in wheat/fallow rotations. However, concerns have been raised about the viability of this important change in land use, within which water supplies is one important concern as large scale biofuels production may place pressure on fresh water supplies and water quality. Despite these achievements on analyzing the water impact of production of selected biofuel crops, investigation of rapeseed production for biofuel has been limited. This study develops assessment informed by inputs from multiple models to evaluate the environmental impacts of water utilization and water quality of the different production options for HRJ production from rapeseed cultivation in North Dakota. The biogeochemical based EPIC model is incorporated to simulate the crop growth that influences the hydrological cycle. Output of the land-use management changes from the EPIC is used to run the blue water footprint analysis using the LCA software SimaPro 8.0. Evapotranspiration data from the EPIC model is incorporated to calculate green water footprint during cultivation. The UNESCO Grey Water Footprint Accounting Tier 1 Guideline is followed to estimate the grey water footprint caused by nutrient leaching and runoff. In addition, inputs for multi-modal transportation logistics were obtained from the Alternative Fuel Transportation Optimization Tool (AFTOT) models in collaboration with the US Department of Transportation Volpe National Transportation Systems Center. Co-product allocation methods were employed in scenario analyses, such as energy allocation and market value allocation, therefore producing a robust analysis of water footprint of the rapeseed HRJ life cycle with compiled life cycle inventory.

Our presentation details the methods, assumptions and initial water footprint results for different scenarios of importance in North Dakota, one of the 10 US states studied in this project, when considering the rapeseed HRJ life cycle. Preliminary results indicate that HRJ fuel derived from rapeseed in rotation with wheat has a much larger water footprint when comparing to the conventional jet fuel. In addition, although it is usually neglected, the indirect blue water footprint caused by the energy usage and the material inputs during HRJ production pathway is worth considering due to its important impact on the overall water footprint. Therefore, policy makers should consider highly the water factor when making decisions on transition to large share of biofuels.

468516 NexSym – a Local Nexus Simulation System

Abstract:

Food, energy and water are essential needs to sustain human life. The complex interactions and interdependencies among the infrastructures and decision making processes involved in the provisioning of such needs have recently generated great debate under the umbrella of the Food-Energy-Water (FEW) nexus (also referred to as Water-Energy-Food nexus). As society becomes increasingly aware that resources available to satisfy the needs of a growing population are finite, there is a call to look at the interactions and start creating more efficient solutions to manage the wide range of nexus scenarios, especially under conditions of climate change, increased urbanisation and waste generation. Furthermore, there is awareness of the importance of maintaining ecosystems to achieve sustainable production while expanding land use and using it more intensively to obtain more food, energy and clean water. The nexus may manifest in unique ways in different localities. Similarly, the global conditions will affect each location differently. Therefore, it is urgent to develop holistic analytical tools that could inform decision making for managing the interdependencies of the FEW nexus at the local scale. This is one of the ways chemical engineers are able to contribute with solutions by applying their skills of modelling processes and managing constraints associated with an interconnected system.

In this work, we present the **Local Nexus Simulation System (NexSym)** that integrates models for technological process components involved in the production of food, energy and clean water as well as waste treatment. Furthermore, the techno-ecological interactions can be simulated by integrating models for ecosystem processes, including agricultural production. The tool then allows the user to build a "flowsheet" of the local system with the component of interest (e.g. bioenergy production) and their interacting local system components (e.g. biomass cultivation, water provisioning). This enables examination of the whole picture of FEW nexus interactions at the local scale, which can then be investigated within a global context. Options for managing the nexus from an engineering perspective can then be devised and further evaluated under scenarios of interest to the user. The **NexSym** is expected to have wide application by engineers working across the nexus system components, planners for urban or rural development, policy makers and decision makers in general. The tool will be illustrated with sample case studies featuring interactions between food, energy, water provisioning and waste treatment to demonstrate the usefulness of such a tool in understanding, managing and creating sustainable solutions to the nexus.

472086 Gibbsian Game Theory for Tragedy of the Commons Problems in Food-Energy-Water Sustainability

Abstract:

Sustainability problems in the food-energy-water nexus often involve "tragedy of the commons" (TOC) problems. TOC challenges are often modeled in game theory using n-player Prisoner's Dilemma (NPD) games and other games. However, classical solutions to game theory problems – here called "Nashian games" – have a number of shortcomings, and do not always conform to experimental data. In this talk we treat strategic games as a set of chemical reactions, and we treat choices as types of reactive molecules. Thus, this model provides new solutions to classical games. By using the usual Gibbsian approach to chemical reaction equilibrium, we are able to employ entropic considerations and minimize a type of Gibbs free energy to find game theory equilibria, overcoming the usual shortcomings in game theory. We solve classical games, including the 2-player PD game and the NPD game, using the "Gibbsian game theory" approach, which accounts for "entropic choices". We then show how treating strategic games as a set of chemical reactions can be used to represent decision situations more accurately than typical games, and we solve for the equilibrium of sets of reactions. Then, we apply Gibbsian solutions to NPD games to problems in sustainability. By quantifying concepts like "trust", "power", and "corruption", we explore opportunities for nudging outcomes of contested decisions in sustainability decisions. When overlapping considerations occur, as in the food-energy-water nexus, the NPD games reveal fundamental political conditions that increase the probabilities for when genuine human solutions can advance. In this talk we describe Gibbsian game solutions for NPD problems, and show how they apply to sustainability problems in the food-energy-water nexus.

461657 Water and Energy Systems in Sustainable City Development: Agent-Based Modelling and Resource Technology Optimization

Abstract:

The grand challenges facing human societies are all closely interconnected with the sustainable provisioning of energy, water and material resources for constantly growing and developing populations, as well as the subsequent processing and management of wastes and pollutions [1]. Urban water and energy systems are expanding and require attention and careful planning to minimize their economic and environmental impact both from an operational and investment perspective. However, there is a lack of a sufficiently systematic approach to address these impacts and achieve the optimal strategies either in the context of research or real-world decision making [2].

Facing the challenges of energy and water systems in urban development, this work describes a decision-making platform combining a comprehensive database, agent-based simulation modelling and resource technology network optimization. The current regional demographics, land-use, infrastructure and economic information is used as input for the initial state of the focused urban system incorporating energy, water and other related resources. Detailed spatio-temporal demand data, obtained by simulating a synthetic population using agent-based models, is further used to plan capacity utilization and expansion by supply-side matching on a cost optimal basis, eventually aiming to explore the optimal design and operational strategies for residential, commercial, industrial, and other sectors in urban energy and water systems, and especially their nexus. Moreover, long-term socio-economic scenarios are addressed in the urban system development through a real-time feedback loop.

The outputs depict, with fine spatial and temporal resolutions, an overall energy-water landscape of the studied urban region (e.g., Greater Accra Metropolitan Area in Ghana as a prototype [3]). The material and energy balance of supply and demand from both macro and micro perspectives are obtained, and further used to propose environmental friendly and cost effective sustainable city investment and development strategies. Several representative use cases are built

under different scenarios to maximize utilization of resources, minimize net waste generation, and improve social welfare to engage local stakeholders. This work is to become a core component of the *resilience.io* platform as an open-source, data-driven, integrated systematic tool gathering social, environmental and economic data to inform urban planning, investment and policy-making for city-regions globally.

References

[1] Bringezu, S. and Bleischwitz, R. eds., 2009. *Sustainable resource management: global trends, visions and policies*. Greenleaf Publishing.

[2] Keirstead, J. and Shah, N., 2013. *Urban energy systems: an integrated approach*. Routledge.

[3] Koppelaar, R., Sule, M.N., Kis, Z., Mensah, F., Wang, X., Triantafyllidis, C., van Dam, K.H., Shah, N., 2016. Urban water and sanitation flows in the Greater Accra Metropolitan Area. *Urban Water Journal* (submitted).

473001 *Sustainable System Synthesis and Analysis using a Novel Sustainability Concept*

Abstract:

Sustainable system analysis and synthesis

using a novel sustainability concept

One of the first definitions for sustainability has been introduced by United Nations in 1987, in a report entitled, Our Common Future (also known as the Brundtland report) which define sustainable developments as "...development that meets the needs of the present without compromising the ability of future generations to meet their own needs." Although this definition captured the essence of the word sustainability but it is very vague in order to be applicable for other organizations. Efforts have been made by researchers to quantify the abstract definitions of sustainability and give it an applicable view. Previous attempts include the ecological footprint (1992, 1996), the use of Fisher information theory (2002), the pressure-state-response model of environmental sustainability (2003), the environmental sustainability index (2005), the sustainability assessment by fuzzy evaluation (SAFE) model (2001-2009), and the AIChE Sustainability Index (2007).

In the current work a novel sustainability concept is put forward, which allows a rigorous quantification of sustainability for a variety of system descriptions. The new concept allows societal and personal input to be incorporated in the sustainability assessment process. Rigorous mathematical criteria are provided for its assessment, involving the global solution of finite dimensional optimization problems. In addition, the proposed concept lends itself to the synthesis of sustainable systems. Case studies for 2-dimensional and 3-dimensional dynamical systems are used to illustrate the introduced new concept, and its applicability to both sustainability analysis and synthesis.

473967 Systematic Approach Towards Establishing Thermodynamic Principles of Sustainable Coupled Industrial-Natural Systems (CINS)

Abstract:

Fundamentals of thermodynamics apply to wide variety of systems at multiple scales such as materials and biogeochemical cycles [1, 2] - ranging from quantum to planetary scale. It will not be an exaggeration to say that systems that violate physical laws of thermodynamics cannot exist as stable systems. This raises two very critical questions for the field of sustainability: a) What are the thermodynamic principles of sustainability science and b) How do we approach hypothesizing, establishing and validation of these principles in a scientifically rigorous way? For the first question, I will qualitatively present some plausible theories of thermodynamics in context of sustainability such as Maximum Entropy principle [3] and non-equilibrium thermodynamics [4]. For the second question, I will present a framework that can be adopted to study thermodynamic properties of coupled industrial-natural systems (CINS) and relating it to approaches for testing the applicability of thermodynamic principles for CINS. The quantitative study will focus on the thermodynamic properties of nutrient flows (Nitrogen) in an industrial ecosystems coupled with regional-ecological nitrogen cycle. Following the course of establishment of thermodynamic laws in other fields, it is suggested that quantitative modeling framework to observe thermodynamic properties over a long span may provide clues to answer question 1 whereas more validation approaches may need to be developed for drawing some general principles.

References

1. D Panagiotaras et al., 2015, "Biogeochemical Cycling of Nutrients and Thermodynamic Aspects", J. Thermodynamic Catalysis, 6:2
2. J.J. Vallino and C.K. Algar, 2016, "The Thermodynamics of Marine Biogeochemical Cycles : Lotka Revisited", Annu. Rev. Mar. Science , 8:333-356
3. R. E. Ulanowicz and B.M. Hannon, 1987, "Life and the production of entropy", Proc. R. Soc. Lond., B 232, 181-192
4. C.G. Chakrabarti and K. Ghosh, 2009, "Non-equilibrium thermodynamics of ecosystems : Entropic analysis of stability and diversity", Ecological Modelling, 220, 1950-1956

474368 Analysis of United Nations Clean Development Mechanism Carbon Emission Reduction Projects from a Life Cycle Assessment Perspective

Abstract:

We analyze, from a life cycle perspective, the effectiveness of the United Nations Clean Development Mechanism (CDM) emission reduction projects implemented in developing countries. Typically, the carbon offset credit allocated to these projects account for avoided direct emissions of GHGs but do not include indirect emissions that occur during upstream and downstream steps (for both avoided as well as the project emissions). We explore, through a case study approach, how these projects perform when life cycle calcualtions are used. Further, we also explore if these proejcts lead to lower (or higher) health and ecosystem impacts (co-benefits). A small but diverse set of renewable energy generation CDM projects are compared on an equivalent bassis to assess any potential problem shifting issues. Results from this study can help policy makers assess effectivenss of such projects in a more holistic manner and fine tune their emission calculation methodologies.

468174 Evolution and Robustness of the Global Agricultural-Phosphorus Trade Network

Abstract:

Containing one of the three essential nutrients for plant growth, phosphatic fertilizers are an integral input to modern agriculture, and will always have a global demand. As the world population continues to grow through mid-century, a significant increase in food production must follow, potentially furthering our current dependence on synthetic fertilizers. While the global resource-criticality of phosphate rock reserves – from which the majority of these fertilizers are derived – may be uncertain, the trade network upon which many countries depend for access to these fertilizers (and their precursors) is a well-documented complex system, subject to an array of physical, economic, and geopolitical constraints. If we are to collectively achieve global food security by 2030, as per the U.N. Sustainable Development Goals, the global security of agricultural nutrients must also be ensured. To formalize questions of access to these nutrients, network analysis provides an attractive quantitative framework with which to examine features of individual relationships and of the underlying structures of complex systems.

The present work focuses on characterizing the structure and understanding the systemic vulnerabilities of the global agricultural phosphorus trade network (GAPTN). We accomplish this by modeling a directed, weighted global network of annual phosphorus-containing commodity exchanges, including all synthetic phosphatic fertilizers and their chemical precursors. All commodities are further normalized to their weight content of elemental phosphorus. As such, nodes are taken to be countries participating in phosphorus trade, and edges are taken to be the aggregate of the directed trade flows of elemental phosphorus between each pair of countries, resulting in the construction of a global agricultural phosphorus trade network (GAPTN). The primary objectives of this study are to (1) characterize the network topology and evolutionary trends of the GAPTN. (2) Elucidate the sets of nodes and edges that are critical to the overall function of the network. (3) Identify those countries most dependent on the network's functioning, and outline engineering and policy measures such that they may ensure their consistent access to this essential resource.

Edges within the GAPTN display mild heterophily, preferentially connecting nodes of dissimilar degree. Based on statistical testing, the distribution of GAPTN node degrees follows a heavy-tail distribution, as does the distribution of in-, out-, and total-strength for 2014. Such distributions indicate that a small subset of all of the participating countries exchange most of the resources that comprise the system, leaving the network robust to random disruptions and highly vulnerable to inopportune ones. The potential for inopportune disruptions stems from both the price volatility of phosphatic fertilizers with respect to energy prices, and the geopolitical tension that surrounds Western Sahara & Morocco, which owns some 75% of the estimated global phosphate rock reserves. Further confirming the implications of the degree distributions, a robustness analysis of randomized versus targeted removal of nodes from the network indicates that some 80% of the network's overall activity halts when some 5% of the nodes do not participate. Results from node-level analyses reveal a small subset of countries to be particularly vulnerable to supply disruptions, even as the overall network has become more redundant and dense from 1990 to 2014. In light of these findings, both international cooperation on phosphorus management and country-level efforts to close local nutrient cycles can aid in improving food security while also minimizing the ecological impacts associated with fertilizer run-off, especially eutrophication.

471802 A Framework for Considering Synergies Between Nature and Engineering from Process to Planetary Scales

Abstract:

Two major requirements for designing sustainable systems include considering all the systems through the entire life cycle to prevent the shifting of impacts to different scales; and to account for the role played by nature while making decisions in order to stay within the carrying capacity of ecosystems to prevent unintended harm or ecological degradation.

The recently developed Process-to-Planet (P2P) modeling framework [1,2] integrates fundamental engineering models and key processes in the upstream supply chain with macroeconomic systems. It improves upon conventional Sustainable Process Design (SPD) methods by incorporating the entire upstream life cycle while solving a process design problem, thus reducing the chances of shifting of environmental impacts across narrow system boundaries. The Techno-Ecological Synergy (TES) framework [3], on the other hand takes into account the carrying capacity of ecological systems by quantifying the demand and supply of ecosystem services, accounting for all the interactions between technological and ecological systems at multiple scales. Including ecosystems within the boundary of traditional engineering design decisions is expected to identify design solutions that aim to keep the impact of engineering activities within the regenerative capacity of ecosystems, depending on the scale of relevance of ecological systems. While the TES framework is the first of its kind to account for interactions between engineered systems and ecological systems, the framework lacks connections with life cycle models and while the P2P considers a comprehensive analysis boundary, it does not consider ecological systems at any scale.

This work introduces the methodology of "Process-to-Planet Techno-Ecological Synergy" (P2P-TES) for designing sustainable systems. The objective behind developing this framework is to make design decisions at the local scale by capturing the entire upstream life cycle boundary. The Structural Path Analysis algorithm (SPA) embedded within the P2P framework helps to distribute the processes or sectors in the system among the different scales. Inclusion of ecosystem services through TES within the P2P framework will expand the solution space and help to discover optimal solutions which enhance both the environmental and economic profitability of the system. In other words, P2P-TES framework is expected to provide "win-win" solutions to the process design problem compared to one where ecosystem services are not included. In addition, accounting for the imbalance between the demand and supply of ecosystem services will enable making design decisions that keep human activities within the carrying capacity of ecological systems. Ecosystem services that will be included in the framework are carbon sequestration, within the national boundary and air quality regulation service within a regional serviceshed.

P2P-TES framework will be applied to two separate case studies which present their own different set of challenges. Successful application to both of them will prove the robustness of the framework as well expose existent deficiencies.

References

[1] Hanes, Rebecca J., and Bhavik R. Bakshi. "Process to planet: A multiscale modeling framework toward sustainable engineering." AIChE Journal 61.10 (2015): 3332-3352.

[2] Hanes, Rebecca J., and Bhavik R. Bakshi. "Sustainable process design by the process to planet framework." AIChE Journal 61.10 (2015): 3320-3331.

[3] Bakshi, Bhavik R., Guy Ziv, and Michael D. Lepech. "Techno-ecological synergy: A framework for sustainable engineering." Environmental science & technology 49.3 (2015): 1752-1760.

AUTHOR INDEX

AIChE
120 Wall Street, FL 23
New York, NY 10005-4020

ISBN 978-1-5108-3422-4

9 781510 834224

CURRAN ASSOCIATES INC.
proceedings.com